It's Alive, But What Is It?

Groups of Living Things	2
One-Celled Organisms	4
Plants with Tubes	6
Plants Without Tubes	8
Fungi	10
Animals with Backbones	12
Animals Without Backbones	14
Glossary	16

Orlando   Austin   New York   San Diego   Toronto   London

Visit *The Learning Site!*
www.harcourtschool.com

Groups of Living Things

Millions of plants and animals live in the world. Each one is an **organism**, or a living thing. Scientists have come up with ways to classify these living things. When you classify, you group things that are alike.

Microscopic organisms were some of the hardest to classify. These are organisms that cannot be seen with the eyes alone. Scientists have to study them under a microscope. All living things have one or more cells. Most microscopic organisms have only one cell. Imagine how tiny they are!

This didinium is eating a paramecium! These organisms can only be seen with a microscope.

Every living thing pictured here is made up of tiny cells.

In some ways, microscopic organisms are like larger living things that have many cells. They need to take in food, water, and oxygen. They also grow and get rid of wastes.

One-celled organisms also have a cell membrane. The cell membrane allows material to pass into the cell. The cell membrane also allows waste material to pass out of the cell.

Some one-celled organisms can make their own food, but others cannot. Some have cell walls and others do not. Although many microscopic organisms have only one cell, they don't all look alike. Scientists came up with names for the different groups of organisms they found.

 MAIN IDEA AND DETAILS What does the cell membrane of a one-celled organism do?

3

One-Celled Organisms

There are two mains groups of organisms that have only one cell. One group is **bacteria**. There are more bacteria on Earth than any other organism. Even a small amount of soil has billions of bacteria.

Bacteria have no nucleus. They do have the same material that is found in a nucleus. However, that material is not inside a separate membrane.

Scientists classify bacteria based on their shape. Bacteria are shaped like rods, balls, or spirals.

Bacteria are part of us, too. They are in and on our bodies. Some bacteria are very helpful. We use bacteria to digest food. We even take in bacteria when we breathe. Other bacteria are harmful. They can make us ill.

There are bacteria in this soil, even though you can't see them.

The other group of one-celled organisms is the **protists**. Unlike bacteria, protists have a nucleus. Some protists can make their own food, while others cannot. Some of them move about, while others do not.

Like plants, protists like algae can make their own food. Algae also put oxygen into the air.

Another kind of protist is the protozoan. Protozoans cannot make food. They have parts that allow them to move around easily. So, they can catch their food. An amoeba is one example of a protozoan.

 MAIN IDEA AND DETAILS How are bacteria classified?

There are many kinds of algae. Algae are protists that are found in watery places.

Plants with Tubes

Our world contains over 200,000 different kinds of plants. Scientists divide them into two main groups to study them more easily.

One group is made up of **vascular** plants. Vascular plants have tubes, or channels. The purpose of these tubes is to carry water and food through the plant so every part gets nutrients.

Three kinds of systems are at work in vascular plants. The root system helps the plant stay in the ground. Roots take in nutrients and water from the ground.

The second system in vascular plants is the stem system. Roots are connected to leaves through the stems. Stems carry water and food from the roots to the leaves. They also keep the plant upright.

The roots, stems, and leaves of this plant are all doing their jobs to keep this plant alive and healthy.

The seeds of this tree form inside the cones.

The seeds of this tree form inside the fruit.

The third system in a vascular plant is the leaves. Leaves do a lot of work. They take in carbon dioxide and water, as well as energy from sunlight.

Leaves give off oxygen, which we use for breathing. They also make food that is carried to the rest of the plant. Some food made by the plant's leaves is stored in the root system.

There are three kinds of vascular plants. Flowering plants have fruits with seeds inside them. Another group is called the cone-bearing plants. They make seeds in cones. Ferns are a third kind of vascular plants. They make spores.

 MAIN IDEA AND DETAILS What do the leaves in a vascular plant do?

7

Plants Without Tubes

Nonvascular plants are plants without tubes or channels. They do not have one route to carry nutrients to the different parts of the plant.

Nonvascular plants have other ways of taking in water and nutrients. A nonvascular plant takes in water on every part of its outer surface. It works the way a sponge works. It takes in water all over. An example of a nonvascular plant is a moss plant.

Nonvascular plants are small. They grow in wet, dark environments. They also grow very close to the soil. This way they can easily take in water.

Just like a sponge, the moss plant soaks up the water it needs.

There is no root system in nonvascular plants. But there are parts like roots that keep the plant in one place. There are also parts like leaves that make food. The food moves from cell to cell. There are no stems in a nonvascular plant to keep it upright or help water travel. Instead, the water also travels from cell to cell.

Nonvascular plants have spores that help them reproduce. The spores travel in water. Different kinds of nonvascular plants make spores in different ways.

Nonvascular plants include mosses, liverworts, and hornworts. Mosses grow on buildings and in between sidewalk cracks. Liverworts and hornworts grow in wet forest land. You can also find them near rivers.

 MAIN IDEA AND DETAILS How does a nonvascular plant take in water?

Moss can grow in many different places, as long as it can get water.

Fungi

Fungi are organisms that absorb food but cannot move about. They look a bit like plants. In fact, fungi used to be classified as plants. They have cell walls, and they grow in an upright direction. Yet fungi are very different from plants.

Scientists now classify fungi in their own group. This is because they are so different from other living things. For example, unlike plants, fungi cannot make their own food.

Fungi take in food from plants and from dead plant materials. They are able to break down nutrients. After the nutrients are broken down, the fungi absorb the nutrients into their cells. Fungi help return nutrients to the soil.

Fast Fact

Chanterelle mushrooms are sometimes compared to apricots in both their color and smell. They are rich in vitamins A and D. Chanterelles have been used in China to aid in vision and problems with breathing.

A mushroom is a form of fungi. Never eat a mushroom that you find in the wild!

> This mold is used to make the medicine called penicillin.

There are many different kinds of fungi. They are divided into groups that include water molds, bread molds, mushrooms, and sac fungi. Yeast is a form of sac fungi.

Fast Fact

In 1928, Sir Alexander Fleming saw that harmful bacteria were destroyed by a mold called *Penicillium notarum*. In the 1940s, scientists made from the mold a powder that was used for a powerful medicine.

Some fungi are safe for humans and can even help us. For example, some mushrooms are safe to eat. Other mushrooms are poisonous. It's important to know the difference. Always check with an adult before eating wild mushrooms.

Yeast is another kind of fungi that is safe and helpful. We use yeast to make bread. It helps bread rise.

 MAIN IDEA AND DETAILS How do fungi get food?

Animals with Backbones

Animals are classified into two main groups. One is called **vertebrates**. These animals have backbones. The bones inside a vertebrate's body make up a skeleton.

Mammals, birds, reptiles, amphibians, and fish are vertebrates. They all have backbones and body systems with many parts. But they look very different.

Mammals are warm-blooded. Their bodies stay the same temperature, even if the air around them is a different temperature. Humans are mammals. All mammals have some kind of hair. Humans have hair. Cats have fur. Porcupines have spines.

Birds are also warm-blooded. Instead of hair or fur, they have feathers that keep them warm and dry. Their bones are light but strong.

These dogs are warm-blooded. Their body temperature stays the same no matter how cold it gets outside.

Snakes need to live in warm climates.

Some reptiles are lizards, snakes, and turtles. Their bodies are the same temperature as the air around them. Reptiles have scales.

Amphibians live both in water and on land. Most amphibians are born in water and have tails. Then they lose their tails and grow lungs and legs. Their bodies are smooth. Amphibians include frogs and salamanders.

Fish have gills so they can take in oxygen underwater. They are also covered with scales.

Amphibians, fish, reptiles, and birds lay eggs. Mammals give birth to live babies. Still, they are all vertebrates.

Fast Fact

The Anaconda snake is very heavy. It can weigh over 500 pounds. Anacondas live on the banks of the rivers and swamps of South America.

MAIN IDEA AND DETAILS What traits do all mammals have?

Animals Without Backbones

Animals that do not have backbones are called **invertebrates**. Without backbones, invertebrates can't stay upright. Many of them live in the ocean. The water supports their bodies.

There are many groups of invertebrates. Jellyfish and sea anemones are sacs. They only have a place that allows them to digest food. Another unusual invertebrate is coral. Did you know coral is actually a group of tiny animals?

Mollusks are another group of invertebrates. They have a foot and a soft body. They also have nerves to carry messages through their bodies. Mollusks include snails and clams.

Fast Fact

The Great Barrier Reef is the largest thing on Earth built by organisms. It is home to almost 2,000 kinds of fish. About 4,000 mollusks and at least 350 hard corals also call it home.

The Great Barrier Reef in Australia is a huge range of coral.

Another group of invertebrates includes worms. Roundworms live in animals. Earthworms live in the ground. Earthworms have a brain and nerves. Did you know each earthworm also has five hearts? There are many kinds of worms.

The largest group of invertebrates is made up of arthropods. Arthropods include insects, crabs, shrimp, and spiders. Arthropods live on land and in water. They have a stiff outer skin called an exoskeleton. As they grow, they shed the exoskeleton and grow a new one.

 MAIN IDEA AND DETAILS Why do many invertebrates live in water?

Like all arthropods, this centipede has jointed legs.

Summary

Scientists classify organisms to study them more easily. Some organisms are too small to see and have only one cell. Other organisms have systems with many parts. Plants are classified as vascular or nonvascular. Animals are classified as vertebrates and invertebrates.

15

Glossary

bacteria (bak•TIR•ee•uh) One of the kingdoms of one-celled living things (4, 5, 11)

fungi (FUHN•jy) Organisms that can't make food and can't move about (10, 11)

invertebrates (in•VER•tuh•brits) The group of animals without backbones (14, 15)

microscopic (my•kruh•SKAW•pik) Too small to be seen with the eyes alone (2, 3)

nonvascular (nahn•VAS•kyuh•ler) Without tubes or channels (8, 9, 15)

organism (AWR•guh•niz•uhm) A living thing (2, 3, 4, 5, 10, 14, 15)

protist (PROH•tist) One of the kingdoms of living things that are one-celled (5)

vascular (VAS•kyuh•ler) Having tubes or channels (6, 7, 15)

vertebrates (VER•tuh•brits) The group of animals with backbones (12, 13, 15)